CONTENTS

Chapter 2: DIY Atmospheric Water Generators	8
Step 3: Set Up Water Collection Containers	12
Chapter 3: Solar Water Condensation	17
Step 5: Solar Heating and Water Collection	21
Chapter 4: Building a Solar Still	25
Step 6: Set Up the Water Collection System	30
Chapter 8: Case Studies and Examples	72

Water from the Air: Simple and Affordable Solutions for Every Home

Introduction

Water scarcity is a global issue affecting communities everywhere, regardless of their economic class.

With the rise of climate change and growing populations, access to fresh water has become a challenge for millions.

Fortunately, the atmosphere around us holds an abundant supply of water, just waiting to be harvested.

This book explores simple, affordable, and sustainable ways to generate water from the air and solar condensation using primitive and inexpensive technology.

The solutions outlined here can be scaled to fit individual needs, whether for personal use, community water supply, or agricultural purposes.

Chapter 1: The Basics of Atmospheric Water Collection

1.1 What is Atmospheric Water Collection?

Atmospheric water collection is a method of extracting water from the

humidity present in the air.

The Earth's atmosphere contains a large amount of water vapor, even in arid environments.

By using simple tools and methods, it's possible to harvest this moisture and convert it into clean, drinkable water.

This process is especially useful in regions with limited access to fresh water but ample humidity.

The concept of atmospheric water generation (AWG) is based on the principle of condensation—where water vapor turns into liquid when cooled or exposed to surfaces cooler than the surrounding air.

1.2 How Does Atmospheric Water Harvesting Work?

The process of collecting water from the air revolves around condensing the vapor into liquid.

This can be done by various methods:

Cooling Surfaces:

Air coolers or metal surfaces in contact with humid air cause water to condense as droplets.

Solar Condensation:

The sun heats the air, causing water to evaporate. By trapping the moisture, we can guide it toward cooler surfaces where it condenses.

Natural Fog Nets:

In fog-heavy areas, large mesh nets can trap moisture from the fog, which drips down into containers.

1.3 Factors That Impact Atmospheric Water Collection

The efficiency of atmospheric water collection depends on several factors:

Humidity Levels:

Higher humidity increases water collection. For instance, coastal areas or tropical regions with consistent humidity above 50% are ideal for AWG.

In drier climates, alternative methods may be needed to increase water yield.

Temperature Differences:

A cooler surface is essential to trigger condensation. The bigger the temperature difference between the air and the surface, the more water you can collect.

Wind Speed:

A gentle breeze can help increase the rate at which air flows over your collection system, thus increasing water harvesting potential.

1.4 Benefits of Atmospheric Water Collection

Low-Cost and Accessible:

This method does not require sophisticated machinery, making it accessible to anyone willing to gather the necessary materials.

Sustainability:

Atmospheric water collection relies on renewable resources, making it eco-friendly and sustainable.

No Need for Fresh Water Sources:

Unlike traditional wells or reservoirs,

AWG doesn't require direct access to groundwater or rivers.

1.5 Limitations of Atmospheric Water Collection

Low Yield in Dry Climates:

In areas with low humidity, atmospheric water collection may not provide sufficient water for all needs.

Scale Limitations:

While simple systems work well for personal use, larger-scale systems require more sophisticated designs to meet community or industrial water demands.

CHAPTER 2: DIY ATMOSPHERIC WATER GENERATORS

2.1 Introduction to DIY Water Generators

Building your own atmospheric water generator (AWG) is an inexpensive and straightforward way to harvest water from the air.

By using common materials and basic engineering principles, you can create a system that suits your water needs.

The design can be easily scaled up or down depending on the amount of water you wish to collect.

This chapter will focus on two simple DIY water generators: one using cooling surfaces and another employing mesh nets for fog collection.

2.2 Materials Required for a Basic Water Generator

To build a simple atmospheric water generator, you'll need:

Metal Sheets or Aluminum Foil: These will act as cooling surfaces for condensation.

Plastic Containers or Buckets:

For collecting the condensed water.

Wooden or Metal Frame:

To support your collection setup.

Mesh Fabric or Plastic Nets:

For fog harvesting (optional for fog-heavy regions).

Plastic Tubing:

For directing the condensed water into containers.

Insulation (optional):

To enhance the cooling of surfaces and improve water yield.

These materials are affordable, widely available, and can be adapted for different setups.

2.3 Step-by-Step Guide to Building a Cooling Surface Water Generator

This design uses the principle of cooling surfaces to condense water from humid air.

Here's how to build a basic atmospheric water generator:

Step 1: Build the Frame

Start by constructing a simple frame out of wood or metal, about 3 to 4 feet tall.

The frame should be sturdy enough to support metal sheets or aluminum foil.

Step 2: Attach the Cooling Surface

Attach metal sheets or large strips of aluminum foil to the frame at an angle.

This slant will allow water droplets to slide down into a collection container.

Position the surface so that it is exposed to the flow of air, ideally in a

shaded area where the surface stays cooler than the surrounding air.

STEP 3: SET UP WATER COLLECTION CONTAINERS

Place a plastic bucket or container at the bottom of the cooling surface.

Water droplets will form on the metal sheet, slide down, and collect in the container.

You can use plastic tubing to direct the water from the bottom of the metal surface into the container, preventing contamination.

Step 4: Maximize Cooling and Condensation

If possible, wrap the back of the metal sheets with an insulating material to enhance cooling.

Alternatively, position the setup near a natural cooling source like a body of water or in an area with consistent shade to help with condensation.

Step 5: Scale the Design

To collect more water, increase the number

of cooling surfaces or build larger frames.

For bigger setups, consider using multiple containers connected by tubing to collect more water at once.

2.4 Step-by-Step Guide to Building a Fog Collector (Fog Nets)

In areas where fog is common, fog nets can efficiently collect water.

Here's a simple method to create a fog collection system:

Step 1: Set Up the Frame

Use a sturdy wooden or metal frame to hold the mesh net.

The frame should be about 6 to 10 feet tall for optimal fog collection.

Position the frame in an open area where fog passes through.

Step 2: Install the Mesh Nets

Attach mesh fabric or fog-catching plastic netting across the frame.

The mesh needs to be fine enough to trap moisture from the fog, but not too dense to block airflow.

Angle the net so that fog flows through it easily.

Step 3: Water Collection System

Place a collection container at the base of the mesh net.

Run plastic tubing from the bottom of the net to the container to direct the water into a clean storage unit.

Step 4: Optimize for More Water Collection

To collect more water, set up multiple frames with mesh nets, spaced a few feet apart.

For best results, position the fog nets perpendicular to the prevailing wind direction, so they intercept the maximum amount of fog.

2.5 Key Considerations for DIY Water Generators

Location:

Choose locations where humidity is consistently high, or where fog is common, to maximize water yield.

Material Durability:

Metal and plastic materials should be corrosion-resistant to ensure longevity.

Maintenance:

Clean the cooling surfaces and collection containers regularly to prevent dirt and debris from contaminating the water.

2.6 Scaling Up the System

To increase water collection, you can:

Use larger cooling surfaces, multiple frames, and larger collection containers.

Combine fog nets and cooling surfaces in the same setup for maximum efficiency.

By adjusting the size and number of components, you can scale the system to suit individual or community water needs.

The simplicity of these designs makes them versatile and cost-effective, allowing anyone to harvest water in humid or foggy environments.

CHAPTER 3: SOLAR WATER CONDENSATION

3.1 How Solar Water Condensation Works

Solar water condensation relies on the sun's heat to evaporate moisture, which is then condensed on a cooler surface, allowing it to be collected as fresh water.

This method has been used for centuries in arid regions where direct access to water is scarce.

A solar still is one of the most common ways to harness this process.

The concept is simple: a transparent cover allows sunlight to heat a water source or damp soil inside the still.

The heat causes water to evaporate and rise, where it cools against the transparent cover, condenses, and is collected.

This method is highly effective, requires

little maintenance, and can be built with inexpensive materials.

3.2 Materials Required for a Solar Still

To build a solar still, you will need:

Plastic Sheeting or Glass Pane:

This will cover the still and trap solar heat.

Collection Basin or Container:

A shallow basin to hold the water or moist soil.

Plastic Tubing or Funnel:

To direct condensed water into a collection container.

Rocks or Weights:

To hold down the cover and create a slope for water to drip.

Digging Tools:

If creating an in-ground still.

Dark-Colored Material (optional):

To enhance evaporation by

absorbing more heat.

These materials are affordable and widely available, making solar stills a practical choice for harvesting water in many environments.

3.3 Step-by-Step Guide to Building a Solar Still

Step 1: Select the Location

Choose an area with plenty of sunlight.

Ideally, this spot should get sunlight for most of the day.

A location with damp soil or close to a water source (like a stream or pond) is ideal for increasing the efficiency of water evaporation.

Step 2: Dig the Pit

Dig a pit that is about 3 to 4 feet wide and 2 feet deep.

The pit size can vary depending on how much water you need to collect. A larger pit will yield more water, but the setup time may increase.

Step 3: Place the Collection Container

Place a shallow basin or container in the center of the pit.

This will collect the water as it condenses and drips down.

If the soil is already damp, you can skip adding any external water.

Otherwise, pour a little water into the pit to help start the condensation process.

Step 4: Cover the Pit

Lay the plastic sheeting or a glass pane over the pit. Make sure it's tightly secured around the edges of the pit using rocks, dirt, or other heavy objects to prevent airflow from escaping.

Place a small stone or weight in the center of the cover so it forms a slight downward slope, directing the condensed water to drip into the collection container.

STEP 5: SOLAR HEATING AND WATER COLLECTION

As the sun heats the air inside the pit, water from the damp soil or the added water will evaporate.

The vapor will rise and condense on the cooler underside of the plastic or glass cover, where it will roll down the slope and drip into the collection basin.

Use plastic tubing or a funnel to direct the water into a separate container if you don't want to open the solar still regularly.

Step 6: Collect the Water

After several hours (typically mid-afternoon or early evening), you can collect the water from the basin.

To collect more water, you can repeat this process over the next several days, especially in regions with ample sunlight.

3.4 Tips for Improving Efficiency

Maximize Heat Absorption:

Place a dark-colored material, such as black plastic or rocks, at the bottom of the pit to absorb more heat from the sun.

This will help speed up the evaporation process.

Increase Moisture:

In extremely dry climates, you may need to add moisture to the pit by pouring in water or even moist plant material to increase the yield.

Use Multiple Stills:

For higher water needs, build multiple solar stills in sunny areas.

Each still can be constructed side by side to collect more water in less time.

Insulation:

In cooler climates, consider insulating the edges of the pit with materials like straw, leaves, or cloth to trap more heat inside.

3.5 Portable Solar Stills

If digging a pit isn't practical, you can create a portable solar still using similar principles. Here's how:

Materials Needed:

A large, clear plastic bottle or a small greenhouse-like structure.

A smaller collection container that can fit inside the bottle.

Dark material to line the inside of the bottle for enhanced heat absorption.

Plastic tubing for directing the condensed water.

Instructions:

Place the collection container inside the bottle or small structure.

Add dark material around the inside to increase heat absorption.

Place the setup in direct sunlight and cover it with a clear plastic cover.

Condensation will form inside, and water can be collected via tubing or directly from the container after several hours.

This design is compact, portable, and ideal for travel or emergency situations.

3.6 Benefits of Solar Condensation

Low-Cost Solution:

Solar stills rely only on the sun's heat and basic materials, making them an inexpensive way to generate water.

No Chemicals Needed:

Unlike other water purification methods, solar stills use no chemicals or filters, making them ideal for long-term sustainability.

Scalable and Adaptable:

Solar condensation can be adapted for individual use or larger community needs by creating multiple stills.

CHAPTER 4: BUILDING A SOLAR STILL

4.1 Overview of the Solar Still

A solar still is one of the simplest and most effective ways to collect water in regions with consistent sunlight.

This method uses the sun's heat to evaporate moisture, which then condenses on a transparent cover and is collected as fresh, drinkable water.

This chapter will guide you through building a solar still, provide tips to enhance its efficiency, and discuss various design adaptations depending on available materials.

4.2 Essential Materials for Building a Solar Still

To build a basic solar still, you'll need the following materials:

Clear Plastic Sheeting or Glass Pane:

Acts as the transparent cover to trap heat and allow condensation.

Shallow Container or Basin:

For collecting the condensed water.

Plastic Tubing or Funnel:

Directs water into a separate collection container.

Dark-Colored Absorbent Material (optional):

Helps increase evaporation by absorbing more sunlight.

Rocks or Weights:

Used to hold down the plastic or glass cover and create a slope for water collection.

Shovel or Digging Tools:

For creating the pit where the still will be placed.

All of these materials are inexpensive and easily accessible, making solar stills a highly practical water collection method.

4.3 Step-by-Step Guide to Building a Solar Still

Step 1: Select the Location

Choose a sunny location that receives sunlight for the majority of the day. Avoid shaded areas to maximize water collection.

Ideally, the location should be near a natural moisture source, such as a pond, damp soil, or even wet vegetation.

Step 2: Dig the Pit

Use a shovel or digging tools to excavate a pit approximately 3 to 4 feet wide and 2 feet deep.

The depth can vary depending on how much water you want to collect and the size of your available materials.

The bottom of the pit should be flat and level so that the collection basin can sit evenly.

Step 3: Prepare the Pit for Evaporation

Place a shallow container or basin in the center of the pit.

This will collect the condensed water as it drips down from the cover.

If the soil is already damp, there's no need to add water.

However, if the area is dry, you can add a small amount of water or moist plant material around the edges of the pit to increase evaporation.

Optionally, line the bottom of the pit with dark-colored absorbent materials, such as black plastic or stones, to help absorb more heat and speed up the evaporation process.

Step 4: Cover the Pit

Lay a clear plastic sheet or glass pane over the top of the pit.

The edges of the cover should extend beyond the pit to prevent air from escaping and to trap heat inside.

Use rocks, dirt, or other weights to secure the edges of the plastic or glass, ensuring it is tightly sealed to maintain the greenhouse effect.

Step 5: Create a Slope

Place a small rock or weight in the

center of the plastic or glass cover to create a downward slope.

This will direct the condensed water toward the middle of the cover, allowing it to drip into the collection basin.

Ensure that the slope is not too steep; a gentle slope works best for directing the water droplets.

STEP 6: SET UP THE WATER COLLECTION SYSTEM

As the water condenses on the underside of the cover, it will drip down into the shallow collection container.

If you prefer not to disturb the still frequently, you can run plastic tubing from the collection basin to an external container.

This way, water can be collected without dismantling the still each time.

4.4 Maximizing Efficiency

To enhance the efficiency of your solar still, consider the following tips:

4.4.1 Increase Heat Absorption

Dark Materials:

Place dark-colored materials such as rocks, soil, or even black fabric inside the pit to absorb more sunlight and increase the rate of evaporation.

Insulation:

Insulate the sides of the pit with materials like straw, grass, or leaves to trap heat and prevent it from dissipating.

This is especially useful in cooler climates or during windy conditions.

4.4.2 Maximize Moisture Availability

Add Moisture:

In extremely dry climates, add additional moisture by pouring water or placing wet plant material in the pit.

This will help increase the amount of water available for evaporation.

Vegetation:

In areas with limited access to water, use vegetation as a moisture source.

Digging the pit near plants or damp soil can enhance the still's efficiency, as plant life tends to retain water.

4.4.3 Use Multiple Stills

If you need more water, consider building multiple solar stills in areas that receive ample sunlight.

Setting up a series of stills allows you to scale up your water collection without requiring more complex designs or equipment.

4.4.4 Adjust the Cover

Material:

Using clear plastic sheeting is generally better than glass in terms of flexibility and heat retention, but glass can last longer if durability is needed.

Angle:

Adjusting the angle of the cover can also affect efficiency.

In regions with low humidity, a slightly steeper angle may help direct water droplets into the collection container faster.

4.5 Testing and Troubleshooting

4.5.1 Test Run

Conduct a test run by setting up the solar still for an entire day, starting early in the morning.

Check the collection basin at the end of the day to measure how much water has been collected.

In dry climates, you may need to leave the still in place for several days to gather enough water, but even in low-humidity environments, the still can produce usable amounts of water.

4.5.2 Common Problems and Solutions

Low Water Yield:

If your still is producing low amounts of water, ensure the pit is properly sealed to prevent heat from escaping.

Adding more moisture or increasing the cover's surface area may also help.

Debris in Collected Water:

Regularly check and clean the collection

basin to ensure the water remains free of dirt or other contaminants.

Using plastic tubing for collection can help avoid exposure to outside elements.

4.6 Variations in Solar Still Designs

While the basic design works well for most applications, there are several variations that can be adapted based on materials and location:

Above-Ground Still:

If digging a pit isn't possible, construct an above-ground version using a frame to hold the plastic cover and place the water collection basin inside.

This version works well in areas with rocky terrain.

Portable Still:

Create a smaller, portable still using a clear plastic container or bottle, a dark interior surface, and a mini collection basin inside.

This is ideal for traveling or emergencies.

Multi-Layer Still:

For larger-scale water production, build a multi-layer still by stacking several frames and covers on top of each other, with each layer acting as a separate evaporation-condensation chamber.

This allows for the collection of water from multiple levels in a single system.

4.7 Advantages of Solar Stills

Low Maintenance:

Solar stills require minimal upkeep once constructed. They rely solely on the sun and do not need electricity, filters, or chemicals.

Sustainable and Eco-Friendly:

Solar stills use renewable energy and are environmentally friendly, making them ideal for remote or off-grid locations.

Adaptability:

Solar stills can be modified and scaled based on available resources and water needs.

Whether for personal use or a larger community setup, this method is highly versatile.

Chapter 5:

Scaling Up for Larger Water Needs

5.1 Introduction to Scaling Up

While basic atmospheric water generators

(AWG) and solar stills are suitable for personal use, larger setups are needed to meet higher water demands, such as for families, communities, or agricultural purposes.

Scaling up requires more advanced designs, efficient material use, and modifications that increase water collection capacity.

In this chapter, we'll explore how to scale up your water collection systems by using modular setups, increasing the surface area for condensation, and integrating multiple units.

We will also discuss ways to store and manage the collected water efficiently.

5.2 Expanding the Collection Area

One of the simplest ways to increase the amount of water collected is to expand the area available for condensation.

Here's how:

5.2.1 Increase Surface Area for Cooling

Larger Cooling Surfaces:

For AWG systems that rely on cooling surfaces (such as metal sheets or aluminum foil), increasing the surface area will capture more water from the air.

You can do this by attaching additional metal sheets to the existing setup or by building a larger frame to support more cooling surfaces.

Vertical Expansion:

Install vertically stacked cooling surfaces that allow air to flow over each layer.

By increasing the number of layers, you effectively double or triple the water collection potential without expanding horizontally.

Multiple Units:

Use multiple AWG systems spaced out over a wide area to increase overall water production.

This modular approach allows for easy maintenance, as each unit operates independently.

5.2.2 Increase Solar Still Size

Larger Pit for Solar Still:

To scale up a solar still, dig a larger pit and use bigger or multiple transparent covers to increase the evaporation area.

A larger solar still can collect significantly more water as it traps more solar heat.

Extend the Cover:

Use a larger piece of clear plastic or multiple sheets to create a larger condensation area.

You can also build a frame to support an extended cover over several pits.

Add More Collection Basins:

If your still is large enough, use multiple shallow basins for water collection to ensure that no condensed water is lost.

5.3 Integrating Multiple Units

When scaling up, integrating multiple AWG or solar still units can be an effective way to meet water needs.

Here's how to do it:

5.3.1 Modular Design

Modular Solar Stills:

Instead of building one massive solar still,

set up several smaller stills in different locations to maximize sunlight exposure.

Each still can be designed independently, but the water they collect can be directed to a single large storage container using tubing.

Multi-Unit AWG System:

For AWG systems, arrange multiple units in rows or grids, allowing for the most efficient use of available space.

This layout ensures that air can flow over all units equally, enhancing water collection efficiency.

5.3.2 Centralized Water Collection

Centralized Storage:

When using multiple water collection units, connect them to a centralized storage system using tubing or pipes.

Each unit will collect water independently but direct it to the same storage container, making management easier and minimizing the risk of water loss.

Gravity-Driven Flow:

To reduce the need for pumping equipment, position your water collection units on higher ground and the storage container on lower ground.

This setup allows gravity to direct the collected water into the storage container.

5.4 Advanced Solar Still Designs

For larger-scale solar stills, you can incorporate more advanced features:

5.4.1 Multi-Layered Solar Still

Vertical Layers:

Build a multi-layered solar still by stacking several evaporation and condensation chambers on top of each other.

Each layer operates independently, allowing multiple surfaces for water to condense and collect.

This setup can produce significantly more water without requiring additional ground space.

Larger Covers and Pits:

Use larger pits and extended plastic covers or glass panes to increase the surface area exposed to sunlight.

Larger stills can condense more water but require more attention to ensure the cover remains airtight and properly sloped.

5.4.2 Solar Desalination Still

Seawater Desalination:

Solar stills can be modified to desalinate seawater, making them suitable for coastal areas.

The basic design remains the same, but you need to ensure the collection basin is sealed and that the salt from the seawater is removed periodically.

Solar-Powered Air Coolers:

Some larger stills use solar energy to power air coolers, which increase the condensation rate by rapidly cooling the air inside the still.

This design is more complex and requires additional resources, but it can produce higher water yields.

5.5 Managing and Storing Collected Water

As you scale up your water collection efforts, managing and storing the water effectively becomes crucial.

Here are some methods to ensure proper water management:

5.5.1 Water Storage Solutions

Large Water Tanks:

Use food-grade plastic or metal tanks for long-term water storage. Ensure that the tanks are properly sealed to prevent contamination from dirt or animals.

Rain Barrels:

If you're collecting water in an area with frequent rain, rain barrels can act as a backup storage system. Place them under your collection units to capture both rainwater and the water harvested from the air.

Gravity-Fed Storage Systems:

To minimize the use of pumps, store water in elevated tanks and use gravity to distribute the water where needed.

5.5.2 Filtering and Purifying the Water

Filtration Systems:

Even though atmospheric water collection generally produces clean water, it's important to filter the collected water to remove any dust, insects, or debris.

Simple gravity-fed filters or homemade sand filters can be used for this purpose.

Boiling or Solar Disinfection:

If you're concerned about microbial contamination, boil the collected water before use.

Alternatively, solar disinfection (placing the water in clear plastic bottles in direct sunlight for 6 hours or more) can help kill pathogens without chemicals.

5.6 Meeting Agricultural Water Needs

In addition to personal and household use, scaled-up atmospheric water collection can be used for small-scale agriculture:

5.6.1 Irrigation Systems

Drip Irrigation:

Collect water in storage tanks and use gravity-fed drip irrigation to water crops efficiently.

Drip irrigation minimizes water wastage and directs water exactly where it's needed.

Hydroponics:

Use atmospheric water collection to support small hydroponic systems.

By using a controlled environment, you can reduce water usage while growing fresh vegetables, herbs, and plants.

5.6.2 Greenhouse Water Collection

Install AWG systems or fog nets inside greenhouses to collect moisture directly from the air and reintroduce it into the growing environment.

This helps maintain humidity levels inside the greenhouse, which in turn reduces the need for external irrigation.

5.7 Examples of Scaled-Up Water Collection Projects

Here are a few examples of how scaled-up atmospheric water collection systems have been successfully implemented:

Fog Nets in Chile:

In the coastal desert of northern Chile, fog nets have been deployed over large areas to collect water for local communities.

These nets capture fog from the Pacific Ocean, producing enough water for drinking and agriculture.

Community Solar Stills:

In regions of India and Africa, large community solar stills have been built to provide clean water for entire villages.

These stills collect and purify water from the air and brackish water sources, offering a sustainable solution for water scarcity.

5.8 Summary of Scaling Up

Scaling up atmospheric water collection involves increasing surface area, using multiple units, and improving water storage and management.

By building larger solar stills, using modular AWG systems, and incorporating centralized water collection, it's possible to meet the needs of larger populations or even agricultural projects.

With the right materials and planning, these systems can be adapted to various climates and environments, providing a sustainable and low-cost solution for water scarcity.

Chapter 6: Maintenance and Optimization

6.1 Importance of Maintenance

Regular maintenance of your atmospheric water generators (AWGs) and solar stills is essential for ensuring consistent water collection, maximizing efficiency, and preventing contamination.

While these systems are relatively simple and low-maintenance, taking a few proactive steps will help extend their lifespan and keep water production at optimal levels.

This chapter will guide you through routine maintenance practices, troubleshooting common issues, and ways to optimize your system to improve water yield.

6.2 Regular Maintenance for Water Collection Systems

6.2.1 Cleaning the Cooling Surfaces (AWGs)

Dust and Debris Removal:

Cooling surfaces such as metal sheets or aluminum foil must remain clean to allow effective condensation.

Periodically clean them with water and a soft cloth to remove dust, dirt, or debris that may accumulate over time.

Inspect for Corrosion:

If you're using metal surfaces, check regularly for signs of rust or corrosion, especially in humid environments.

Replace or treat corroded surfaces to prevent contamination of the collected water.

Ensure Proper Airflow:

For AWGs, make sure air can flow freely over the cooling surfaces.

Remove any obstacles that might block airflow and ensure that the surface angle is correctly positioned to catch humid air.

6.2.2 Maintaining Solar Stills

Clean the Plastic or Glass Cover:

Over time, the transparent cover of the solar still may become cloudy or dirty, reducing the amount of sunlight that enters the pit.

Clean the cover regularly with water and a mild detergent to keep it clear.

Check for Leaks:

Inspect the edges of the plastic sheeting or glass cover to ensure that it is tightly sealed around the pit.

If there are any gaps, seal them with rocks, dirt, or an adhesive to maintain the airtight environment necessary for condensation.

Clean the Collection Basin:

Water collected in the basin may gather dirt or algae if left unattended.

Regularly clean the basin with soap and

water to prevent contamination.

6.2.3 Water Storage Maintenance

Storage Tank Cleaning:

If you're storing collected water in tanks or barrels, clean them at least once every few months to prevent the buildup of sediment, algae, or bacteria.

Check for Leaks:

Regularly inspect water storage containers for leaks or cracks. Use waterproof sealants to repair any damage and prevent water loss.

Filter and Purify Water:

Even with proper maintenance, it's important to filter and purify the collected water before consumption.

Gravity-fed filters, sand filters, or boiling the water will ensure it's safe to drink.

6.3 Troubleshooting Common Issues

6.3.1 Low Water Yield

If your system isn't producing enough

water, consider the following solutions:

Increase Surface Area:

For AWGs, add more cooling surfaces or extend the size of your metal sheets.

For solar stills, use larger covers to capture more sunlight.

Check Humidity Levels:

If the air is particularly dry, your AWG may not collect much water.

You can increase water yield by placing the system in a more humid location or adding moist materials to your solar still.

Adjust the Angle of the Surface or Cover:

Ensure that the angle of your cooling surface or solar still cover is correctly positioned to maximize condensation and direct water into the collection basin.

6.3.2 Water Contamination

If the collected water looks dirty or contains debris, you may need to take the following steps:

Clean the Collection Surfaces:

Dirt or organic matter on the cooling surface or the inside of the solar still can cause contamination.

Ensure these surfaces are clean at all times.

Filter the Water: Always filter the collected water to remove any remaining particles.

Simple cloth or mesh filters can be used for larger debris, while finer filters can remove smaller contaminants.

6.3.3 Leaks and Heat Loss

Leaks in your system can cause a drop in efficiency and reduce water yield:

Seal Any Gaps or Cracks:

Regularly inspect the seams of your solar

still's cover or the connections between the cooling surface and collection containers.

Use adhesive, rocks, or additional plastic sheeting to seal any gaps.

Insulate Your System:

In cooler climates, you may experience heat loss in your solar still.

Insulate the pit or collection area using straw, leaves, or other natural materials to retain heat and improve evaporation rates.

6.4 Optimizing Water Collection Efficiency

6.4.1 Maximizing Condensation

Increase Humidity:

Position your system in areas with higher humidity, such as near bodies of water or damp ground.

If necessary, add wet vegetation or water to the system to increase evaporation and condensation.

Use Dark Materials:

For solar stills, line the bottom of the pit with dark-colored materials like black plastic or stones.

These absorb more sunlight, which increases the evaporation rate and leads to higher water yields.

Cooling Surfaces for AWGs:

If using a cooling surface system, place it in a shaded area to keep the metal cool.

This ensures that the warm, humid air condenses quickly when it comes into contact with the cooler surface.

6.4.2 Expanding Water Collection Capacity

Modular System:

If you require more water, expand your system by adding more AWG units or solar stills.

A modular design allows you to scale up without significantly increasing the complexity of maintenance.

Automate Water Collection:

Use plastic tubing to direct water from multiple units into a single storage container.

This allows for continuous collection without needing to manually empty each system.

6.4.3 Environmental Considerations

Positioning and Location: For AWGs, position your system where airflow is constant.

Avoid areas where the system might be blocked by structures or dense vegetation.

For solar stills, ensure that they receive direct sunlight for the majority of the day.

Seasonal Adjustments:

As seasons change, so do temperature and humidity levels. In cooler months, add insulation around the solar still and adjust the angle of the cover to capture more sunlight.

In hot, dry months, increase water availability by moistening the still's surroundings or adding a water source.

6.5 Long-Term Efficiency

6.5.1 Routine Inspections

Conduct regular inspections of all

parts of your system, from the cooling surfaces to the collection containers.

Early detection of issues such as leaks, cracks, or material wear will prevent more significant problems down the road.

Maintain a schedule for cleaning and inspecting the system to ensure long-term efficiency.

6.5.2 Material Upgrades

Durable Materials:

As your system ages, consider upgrading to more durable materials such as stainless-steel cooling surfaces or glass panes for the solar still.

These materials may be more expensive initially, but they reduce the need for frequent replacements.

Weather-Resistant Covers:

For solar stills, replace plastic sheeting every few years or sooner if you notice wear and tear.

Weather-resistant materials can extend the lifespan of your setup and increase water yield by maintaining an airtight seal.

6.6 Adapting to Climate and Environment

6.6.1 Dry and Arid Climates

In very dry environments, AWGs may not produce as much water due to low humidity. In these cases, consider using solar stills that can utilize water from other sources like brackish water or seawater.

Supplement the AWG by positioning it near areas where fog or dew is more likely to form, such as valleys or coastal regions.

6.6.2 Humid and Tropical Climates

High-humidity environments are ideal for AWGs, which can collect large amounts of water.

In these climates, focus on expanding the surface area of your AWG systems.

Ensure that solar stills are placed in areas with good airflow to prevent overheating, and consider using shade cloth to avoid excess heat, which can reduce the lifespan of the system.

6.7 Summary of Maintenance and Optimization

Routine maintenance and thoughtful

optimization are essential for keeping your water collection systems functioning efficiently.

Cleaning cooling surfaces, checking for leaks, and ensuring proper airflow or sunlight exposure will improve water yield and prevent contamination.

By expanding the collection area, integrating multiple units, and optimizing for specific environmental conditions, your system will continue to provide reliable water harvesting for years to come.

Chapter 7: Sustainability and Environmental Impact

7.1 Introduction to Sustainability in Water Collection

Water is one of the most vital resources on Earth, but it is also one of the most strained.

Traditional methods of water extraction—such as groundwater wells, river diversion, and large-scale desalination plants—

can have significant environmental impacts, including habitat disruption, depletion of natural resources, and the consumption of large amounts of energy.

Atmospheric water collection and solar condensation offer an alternative solution that is environmentally sustainable, energy-efficient, and scalable for various needs.

This chapter will explore the environmental benefits of these systems, how they reduce resource stress, and their long-term sustainability.

7.2 How Atmospheric Water Collection Supports Sustainability

7.2.1 Renewable Water Source

Harnessing Water from Air:

Atmospheric water collection uses the natural moisture present in the air, a renewable resource that replenishes itself through the water cycle.

Unlike groundwater or surface water sources, atmospheric water does not deplete finite reserves, making it a sustainable solution for water-scarce regions.

No Dependence on Large Infrastructure:

Unlike desalination plants or large-scale water treatment facilities, atmospheric water generators (AWGs) and solar stills do not require heavy infrastructure.

This makes them ideal for areas where building large infrastructure may not be feasible or environmentally friendly.

7.2.2 Reducing Pressure on Natural Water Sources

Alleviating Groundwater Depletion:

In many parts of the world, over-extraction of groundwater has led to critical shortages, land subsidence, and ecological damage.

By collecting water directly from the air, AWGs reduce the need for extracting groundwater, allowing aquifers and natural reservoirs to recover.

Preserving Ecosystems:

Diverting rivers and streams for human use can disrupt ecosystems, leading to the loss of plant and animal habitats.

By utilizing atmospheric water instead of diverting natural water bodies, ecosystems remain intact and more balanced.

7.2.3 Low Environmental Footprint

Minimal Energy Use:

Atmospheric water collection systems, particularly solar stills, rely on renewable solar energy.

This eliminates the need for fossil fuels or electricity, reducing greenhouse gas emissions and minimizing the carbon footprint of water collection.

No Chemical Waste:

Unlike industrial water treatment processes, which may involve chemicals or heavy filtration systems, AWGs and solar stills do not produce harmful byproducts.

The water collected is free of chemical additives, and the system itself generates no waste.

7.3 Solar Condensation as a Green Technology

Solar condensation offers a highly sustainable and environmentally friendly way to collect water.

Here's why:

7.3.1 Utilizing Renewable Solar Energy

Zero-Energy Input:

Solar stills rely solely on the sun's energy to evaporate and condense water.

This makes them a zero-energy technology, reducing the need for electricity and lowering overall energy consumption.

Long-Term Use of Natural Resources:

Sunlight is a nearly infinite resource.

Using solar energy to power condensation ensures that the water collection process can be sustained indefinitely without putting a strain on other energy sources.

7.3.2 Water Purification with No Environmental Impact

Natural Water Purification:

Solar condensation also purifies water by distillation, removing salts, bacteria, and other contaminants.

This process leaves no waste behind and does not require the use of harmful chemicals or filtration systems.

Ideal for Remote or Off-Grid Locations:

Since solar stills do not require any external power source, they are an ideal solution for remote, off-grid locations where conventional water treatment infrastructure is not available.

This makes them highly adaptable and accessible for sustainable living.

7.4 Reducing Resource Consumption

7.4.1 Reducing Energy Use

Energy Efficiency in AWGs:

While some advanced AWG systems may require energy to power cooling units, there are low-energy alternatives that rely on natural cooling methods, such as wind and temperature differentials.

These systems use far less energy than traditional water treatment or desalination plants.

Solar-Powered Alternatives:

Many AWG systems can be adapted to run on solar power, making them even more sustainable by reducing their dependence on non-renewable energy sources.

7.4.2 Lowering Water Waste

Efficient Water Collection:

Both solar stills and AWGs collect water with minimal waste.

The condensation process captures nearly all the moisture in the air or water source, making these systems highly efficient.

Localized Water Collection:

By producing water where it is needed, AWGs and solar stills reduce the need for long-distance water transportation, which is often associated with significant water loss due to leaks and evaporation in pipelines.

7.5 Sustainability in Arid and Water-Scarce Regions

In regions where water scarcity is a critical issue, atmospheric water collection can provide a sustainable solution.

Here's how:

7.5.1 Addressing Desertification

Preventing Desert Spread:

In many areas, over-extraction of groundwater leads to desertification, as plants and soil are deprived of moisture.

By harvesting water from the air instead of the ground, AWGs can help slow or prevent the spread of deserts and support sustainable agriculture.

Reforestation and Crop Support:

In areas prone to desertification, AWGs can supply water for irrigation and reforestation projects.

This supports the restoration of ecosystems and encourages sustainable agriculture in arid environments.

7.5.2 Meeting Community Needs

Water Security for Vulnerable Communities: In rural or underserved communities, access to clean water is often a significant challenge.

By setting up AWG systems and solar stills, communities can become more self-sufficient, reducing their reliance on external water sources.

Reducing Water Transportation Costs:

Transporting water over long distances is both costly and environmentally damaging.

By producing water locally, AWGs and solar stills eliminate the need for costly infrastructure, fuel consumption, and water loss during transportation.

7.6 Adaptability for Climate Change

As climate change intensifies, regions around the world are experiencing shifts in water availability.

Atmospheric water collection and solar condensation are adaptable solutions that can mitigate the effects of climate change:

7.6.1 Coping with Changing Rainfall Patterns

Alternative Water Sources in Droughts:

In areas affected by prolonged droughts, traditional water sources such as rivers, lakes, and reservoirs may dry up.

AWGs and solar stills can continue to produce water, even in areas with irregular rainfall, ensuring a stable water supply during droughts.

Capturing Water from Fog and Dew:

As climate change alters weather

patterns, some regions may see increases in fog and dew formation.

AWGs can capitalize on these conditions by collecting moisture from the air, offering a sustainable water source even when rainfall is scarce.

7.6.2 Supporting Climate-Resilient Communities

Localized Water Systems:

Communities can become more resilient to the impacts of climate change by developing localized water systems that are not dependent on external sources.

This reduces vulnerability to water shortages caused by natural disasters, infrastructure breakdowns, or political instability.

Sustainable Agriculture in Changing Climates:

In agricultural regions facing unpredictable rainfall patterns due to climate change, AWGs and solar condensation systems can help provide a reliable source of water for crops and livestock, supporting food security and sustainable farming practices.

7.7 Environmental Impact of Scaling Water Collection Systems

As discussed in previous chapters, scaling up atmospheric water collection systems can meet larger water needs without significant environmental impact.

Here's how:

7.7.1 Scaling Without Resource Strain

Sustainable Expansion:

Expanding water collection systems is possible without increasing energy consumption or environmental damage.

Modular AWG systems and solar stills can be scaled by adding more units, allowing water collection to grow in a sustainable manner.

Community and Agricultural Use:

In addition to personal use, scaling up these

systems can support entire communities, agricultural operations, or businesses without relying on large-scale infrastructure projects that may harm the environment.

7.7.2 Lowering Long-Term Environmental Costs

Durable, Low-Waste Materials:

While AWGs and solar stills can be built with basic materials, more durable systems made with corrosion-resistant metals or weather-proof plastics will have a longer lifespan, reducing the need for frequent replacements and lowering long-term waste.

Minimal Land Use:

Both AWGs and solar stills require minimal land use compared to conventional water systems.

This makes them less disruptive to the natural landscape and ideal for areas where land availability is limited.

7.8 Summary of Sustainability and Environmental Impact

Atmospheric water collection and

solar condensation offer sustainable, environmentally friendly solutions to the global water crisis.

By reducing energy consumption, alleviating pressure on natural water sources, and providing adaptable systems for arid and climate-affected regions, these methods promote water security while minimizing environmental harm.

As climate change continues to challenge water availability, AWGs and solar stills present scalable, renewable solutions that support both communities and ecosystems.

CHAPTER 8: CASE STUDIES AND EXAMPLES

8.1 Example 1: DIY Atmospheric Water Generator (Basic Cooling Surface)

Overview:

This is a simple and cost-effective design using metal sheets or aluminum foil as a cooling surface to condense moisture from the air.

It's ideal for areas with moderate humidity and can be scaled up by increasing the surface area.

How It Works:

The cooling surface is exposed to the air, where water vapor condenses on the cooler metal sheet.

The water droplets roll down into a collection basin.

Materials Used:

Metal sheets or aluminum foil

A plastic container for water collection

Wooden or metal frame to hold the surface

Plastic tubing for directing water into the container

Image Description:

An image of a simple atmospheric water generator with a metal sheet angled on a frame, with water droplets forming and flowing down into a collection container below.

8.2 Example 2: Fog Nets for Water Collection

Overview:

Fog nets are large mesh screens designed to trap moisture from the fog in fog-heavy regions.

This system is effective in coastal or mountainous areas where fog is prevalent.

How It Works:

The fog passes through the fine mesh, where moisture is trapped as droplets.

The water drips into a collection basin or is funneled into a storage container.

Materials Used:

Mesh fabric or plastic netting

Wooden or metal frame to support the net

Collection basin or plastic tubing to direct water.

Image Description:

An image of a large fog net stretched across a metal frame, with fog rolling through the mesh. Water droplets are visible on the net,

dripping into a container at the bottom.

8.3 Example 3: Solar Still (Basic Ground Design)

Overview:

This solar still is a simple, in-ground setup designed to collect water by evaporating moisture from damp soil or a small water source using the sun's heat.

How It Works:

A pit is dug in the ground, with a container placed in the center.

Clear plastic or a glass pane is used to cover the pit, trapping heat.

Water evaporates, condenses on the underside of the plastic, and drips into the container.

Materials Used:

Clear plastic sheeting or a glass pane

Shallow collection basin
Rocks to hold down the plastic sheeting
Dark material for heat absorption (optional)

Image Description:

An image of a solar still pit covered by clear plastic sheeting, with a small rock creating a slope to direct the water into a collection basin in the center.

8.4 Example 4: Portable Solar Still (Above-Ground)

Overview:

A portable version of the solar still, this above-ground design uses a clear plastic container or bottle to capture water via solar condensation, making it ideal for travel or emergency use.

How It Works:

Water is placed inside a clear container or bottle, which is left in direct sunlight.

The sun heats the water, which evaporates and condenses on the inner surface of the container.

Condensed water drips into a collection section.

Materials Used:

Clear plastic container or large bottle

Dark-colored material inside for heat absorption

A small collection basin inside the container

Image Description:

An image of a portable solar still, with a

clear plastic bottle positioned in the sun.

The inner surface shows condensation forming, with water dripping into a small basin inside.

8.5 Example 5: Scaled-Up Solar Still (Multi-Layer)

Overview:

This is a multi-layered solar still designed to maximize water collection by stacking multiple layers of evaporation and condensation chambers. It's ideal for scaling up water production in sunny regions.

How It Works:

Multiple evaporation layers are stacked, each trapping solar heat and evaporating water.

Each layer has its own condensation surface, where water drips down into separate collection containers.

Materials Used:

Multiple transparent covers (plastic or glass)

A multi-tiered frame for support

Collection basins at each level

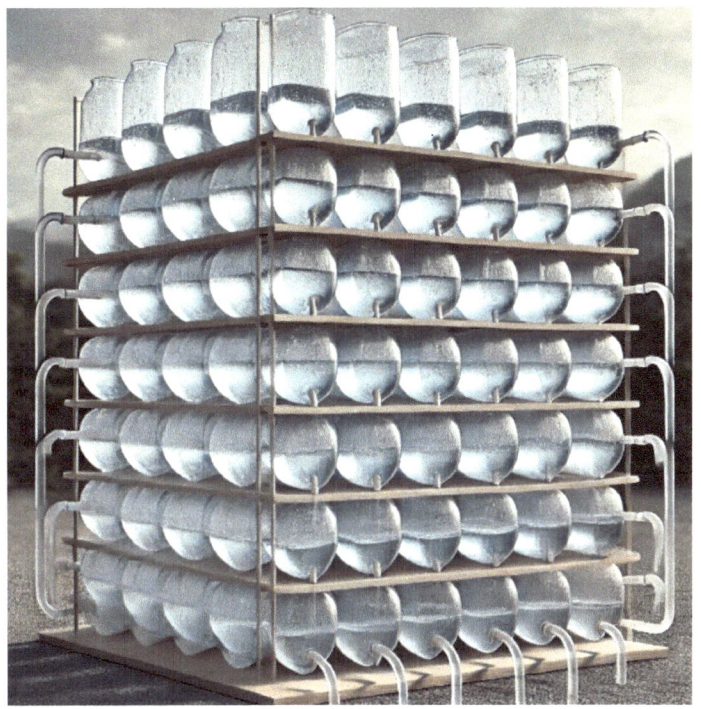

Image Description:

An image of a multi-layer solar still, with several transparent plastic covers stacked vertically. Each layer is shown collecting water, with tubes directing the water into collection containers.

Image of a Pitcher of Water Overflowing: A pitcher of water is overflowing, with water spilling over the sides and splashing around the base on the table.

WATER FROM THE AIR

Image of a Rain Barrel: A rain barrel is placed outdoors, connected to a gutter downspout, collecting water in a lush garden environment.

Image of a Large Pond: A calm pond of water surrounded by greenery, reflecting the sky and trees above, showcasing the serenity of a natural water source.

Image of an AWG Made from a Swimming Pool: A repurposed swimming pool acting as an atmospheric water generator, with a transparent cover capturing condensation and directing water into a collection system.

Image of a Tabletop Solar Still: A small solar still on a table, with a shallow bowl, transparent plastic sheet, and a collection basin for condensed water droplets.

Thank you for sharing my love of practical and low cost sustainable ecological hydrology for the enviroment.

www.ingramcontent.com/pod-product-compliance
Lightning Source LLC
Chambersburg PA
CBHW040223220526
45473CB00001B/94